Green Chemistry in Undergraduate Practical Courses

Jeff Hardy and James H. Clark

Marjorie Cutter Scholarship

The Royal Society of Chemistry 2006

RSC | Advancing the Chemical Sciences

Green Chemistry in Undergraduate Practical Courses

Written by Jeff Hardy and James H. Clark

Edited by Josephine Tunney

Designed by Terry Roberts

Published and distributed by Royal Society of Chemistry

Printed by Royal Society of Chemistry

For further information on other educational activities undertaken by the Royal Society of Chemistry contact:

Education Department
Royal Society of Chemistry
Burlington House
Piccadilly
London
W1J 0BA
Email education@rsc.org
Telephone 020 7440 3344

Information on other Royal Society of Chemistry activities can be found on its website at www.rsc.org

ISBN 0-85404-278-4
British Library Cataloguing in Publication Data.
A catalogue for this book is available from the British Library.

Contents

Introduction

Chemistry undergraduate practical experiments are the primary route to providing students with the tools and techniques required for a career in research or in the chemical industry. However, it is perceived that practical courses in UK universities are rather outdated because they rely upon experiments developed in the past that have not been updated to keep pace with chemistry research and modern industrial practice.

This is resulting in chemistry students leaving university without having hands-on experience of modern chemical techniques such as combinatorial chemistry, micro-analysis, enzymatic catalysis and green chemistry. It is important to consider that the laboratory methods we teach students are the off-the-shelf methods that they will apply during their research/industrial careers and therefore it is our responsibility to ensure that they are equipped with modern examples of best practice in practical chemistry. To partly address this need the Green Chemistry Group (GCG) at the University of York have developed four undergraduate practical experiments based on the principles and practice of green and sustainable chemistry. The practicals have been developed and trialled by Dr Jeff Hardy* over the past four years and represent examples of clean synthesis, renewable feedstocks, heterogeneous catalysis, continuous reactors and phase transfer catalysis. The experiments cover all levels of chemistry undergraduate studies and are also applicable to Masters level students. They have all† been successfully trialled in the teaching laboratories at the University of York by second year undergraduate students (experiments 1 and 2) and Masters level students (experiments 1, 2 and 4). The experiments have been reviewed by Dr Neil Robertson of the Department of Chemistry at the University of Edinburgh and Professor David Knight of the Department of Chemistry at Cardiff University and their positive feedback has resulted in constructive improvements being made to the experiments and scripts.

* *Acknowledgement is also given to the hard and successful work put in by Iria Rodriguez, David Jackson and Owain Samuel.*

† *With the exception of practical 3 which has been trialled in research laboratories by a student but has not yet been trialled in undergraduate laboratories.*

Summary of Cutter Bequest practicals developed by the GCG

Practical sessions	Target students
1. Synthesis of tetrahydrofuran by reactive distillation	1st year undergraduates
2. Alkylation of *p*-xylene with benzyl chloride using homogeneous and heterogeneous Lewis acid catalysts	2nd and 3rd year undergraduates
3. Clean and efficient synthesis of 4-aminobenzoic acid from 4-nitrotoluene	2nd and 3rd year undergraduates
4. Extraction and conversion of limonene to terephthalic acid via *p*-cymene	3rd year undergraduates and Masters students

How to use this resource

This guide contains a summary of the experiments and tutor notes based on the experience of trialling the practicals at the University of York. Each experiment is also accompanied by suggestions for extending the practical work.

Student manuscripts containing detailed experimental information can be found on the accompanying CD-ROM.

Safety

The experiments contained in this guide are designed to be carried out by undergraduate students in the presence of experienced and suitably qualified laboratory demonstrators.

Reasonable care has been taken to ensure that the procedures associated with each experiment are reliable. Even though every reasonable effort has been made by the author to ensure the accuracy of the information contained within this book, the RSC makes no representation, express or implied, with regard to the accuracy of the information contained in this book and is not responsible for any error found in the book.

The person(s) using this information must follow all the customary appropriate procedures for the safe storage, preparation, handling and disposal of any chemicals or other potentially dangerous material mentioned in these pages. The RSC recommends that the experiments in this book are carried out in accordance with the full instructions contained within, and encourages reference to original sources. To the extent permitted by law, the RSC does not accept any liability for the use or misuse of the information in this book. In particular, the RSC accepts no liability for any loss or damage resulting from the failure to follow instructions for the procedures described in the book, or from the incorrect storage, preparation, handling or disposal of the recommended chemicals.

Introduction to Green Chemistry metrics

Green chemistry metrics are analysis tools to quantify the efficiency and environmental impact of chemical processes. In theory such tools could be used to drive decisions in industry, government and academic research towards sustainable practices. For a good example of green chemistry metrics see the review by Curzons *et al* [1]. In the practicals, several green chemistry metrics can be applied in interpreting the results. The three that can be commonly used are % atom economy (%AE), E-factor and % mass productivity (%MP) and these are described briefly below.

% AE is a metric designed to show the atomic efficiency of a reaction.
The calculation is:

%AE = (m.w. of product/m.w. of all reactants used in reaction) x 100 (1)

%AE basically measures how much of the reactants end up in the final product. It does not, however, take into account reaction yield, molar excess of reactants, solvent or reagents (such as stoichiometric oxidants).

E-factor is a measure of the total waste produced per kg of product and as such represents the efficiency of a process in terms of mass usage but not in terms of energy. The calculation is:

E-factor = Total kg waste/kg product (2)

Thus E-factor is a useful measure of the "wastefulness" of a given chemical process and can be a useful tool for highlighting processes of concern.

%MP and mass intensity (MI) is a metric that takes into account % yield, reaction stoichiometry, solvent and reagents used. It is a calculated as follows:

%MP = (1/Mass Intensity) x 100 (3)

MI = Total mass used in a process or process step (kg)/total mass of product (4)

Water (such as cooling water or water for reaction quenching) is often not included in %MP calculations as it tends to completely dominate the calculation.

Introduction to the practicals

1 Synthesis of tetrahydrofuran (THF) by reactive distillation

This simple practical is aimed at first or second year undergraduate students. It is an example of bulk chemicals synthesis and separation carried out in a clean and efficient manner (Figure 1). The equipment required is a simple distillation apparatus. The yield is >90% and the catalyst can be reused several times. The practical can be completed in one four hour laboratory session. There is also the potential to expand it into a mini-project.

Practical aims

The aim of this practical is to give students hands-on experience in several fields, including:

- Synthesis of THF
- Practical experience of reactive distillation
- Using a heterogeneous catalysis
- Quantitative analysis
- Carrying out a reaction with a high % atom economy and a low E-factor
- Green metric analysis of reaction systems

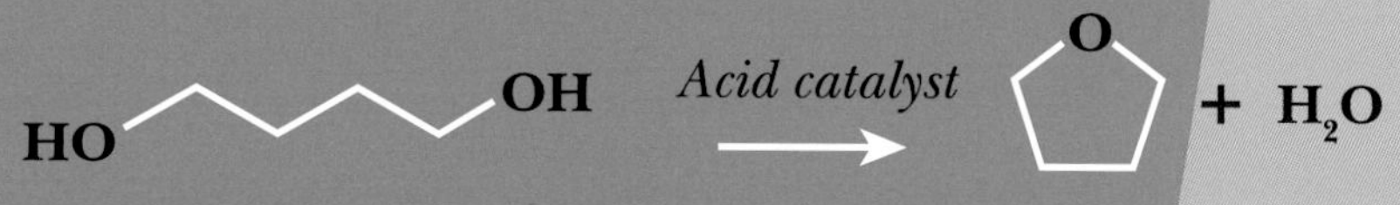

Figure 1: The acid catalysed synthesis of THF from 1,4-butanediol

Learning and discussion points arising from experiment

Students carrying out this practical are generally astounded as to how clean and simple chemical reactions can be. This reaction is an example of a very atom efficient reaction (AE % = 82%, E-factor <1) and is also an example of a continuous process with integrated product separation. The use of a heterogeneous acid catalyst introduces students to the concept that supported catalysts can be as effective as homogeneous catalysts, but are also readily separable from the reaction system unlike the homogeneous analogues. The analysis and purification of the THF product raises a number of interesting learning and discussion points for the students, including:

- How to determine the water content of THF
- The concept of azeotropic mixtures and the problems with purification by distillation or other means
- How to dry "wet" solvents

Description of the experiment

In this practical the students synthesise THF from 1,4-butanediol (1,4-BDO) using an acid catalyst, thus mimicking the current industrial process. The experiment has worked successfully with two groups of students (22 second year undergraduates in the first instance and 13 Masters students in the second instance). The reaction is carried out as a reactive distillation using standard distillation apparatus. The catalysts that have been tested are concentrated sulphuric acid and sulfated zirconia (a heterogeneous acid catalyst). It is expected that other strong Brønsted acid catalysts could also be used to catalyse this reaction although no others have been tested. The starting materials are reasonably cheap (although pure zirconia can be quite expensive). In order for the reaction to proceed, the 1,4-BDO must be heated (by heating mantle) to temperatures in excess of 150°C. Because of the high temperature, and because of the volatile nature of the THF product, it is recommended that the experiment be carried out in a fume-cupboard. The preparation of THF can comfortably be carried out in a four hour laboratory session. If the students choose to dry and analyse the water content of the THF then a second four hour laboratory session on a consecutive day may be required as the drying steps can take 24 hours. This practical is flexible and has the potential to be expanded into several mini-projects.

Possible extensions include:

1. **The THF can be dried.** In the practical carried out at York the students were asked to suggest drying methods and these included distillation, sodium metal (to be discouraged as the water content is high), typical drying agents (such as magnesium sulfate), molecular sieves and salt addition (such as NaCl, which when added in large concentrations causes phase separation). In general these methods were found to remove a significant portion of the water from the THF, but did not dry it completely (as seen by refractive index and by the failure of the Grignard reaction).

2. **Determination of water content.** The students at York have carried out determination of water content by density and by refractive index. The method for refractive index[i] proved to be more accurate than that based on density. The students can also record an infrared spectrum of their "wet" and "dry" THF to examine differences in the –OH stretching and bending regions.

3. **Re-use the sulfated zirconia catalyst** until it becomes inactive (approximately 5 runs). Work out turn over number (TON).

4. **Synthesise THF using different acid catalysts.** Sulfuric acid works but leaves a dark residue in the reaction flask indicating side reactions (such as polymerisation). No other acids have been tried.

5. **Use the THF as a solvent either in a reaction or in chromatographic separation.** The THF must be dried first (unless water is not an issue). At York the students dried THF and used it as a solvent in a Grignard reaction to synthesise crystal violet and malachite green (based on a literature procedure). The reaction failed as the THF was still wet.

6. **Making polyTHF** or using THF as a substrate in another appropriate reaction.

7. **Synthesise THF from furfural** or from a source of biomass high in hemi-cellulose.

[i] *The group as a whole will have to run standard calibration samples. Each pair will be assigned a sample to prepare and measure. The calibration samples are mixtures of THF and water in the following proportions (THF by weight %):*

% THF by weight - 0, 15, 30, 50, 60, 70, 80, 90, 95, 100

For example 1g of the 15 wt % solution would consist of 0.15g THF and 0.85g of deionised water. Make up the calibration standards in ***accurately approximate 5g solutions.***

The refractive index of each calibration standard should be measured and the results made available to all pairs. Each pair should also measure the refractive index of their synthesised THF samples (before and after drying). Calculate the % water (by weight) in the sample from the calibration curve.

2 Alkylation of *p*-xylene with benzyl chloride using homogeneous and heterogeneous Lewis acid catalysts

This practical is aimed at chemistry undergraduate students in their second or third year. It critically examines the difference between the homogeneous (aluminium chloride) and heterogeneous (zinc chloride supported on K100 silica) catalysis of a Friedel-Crafts alkylation reaction (Figure 2). The full practical can be comfortably completed in two five hour practical sessions.

Practical aims

- To carry out a Friedel-Crafts alkylation reaction
- To compare the homogeneous and heterogeneous catalysis
- To assess quantitatively the amount of waste generated during this reaction
- To critically assess the pros and cons of each reaction from a Green Chemistry perspective

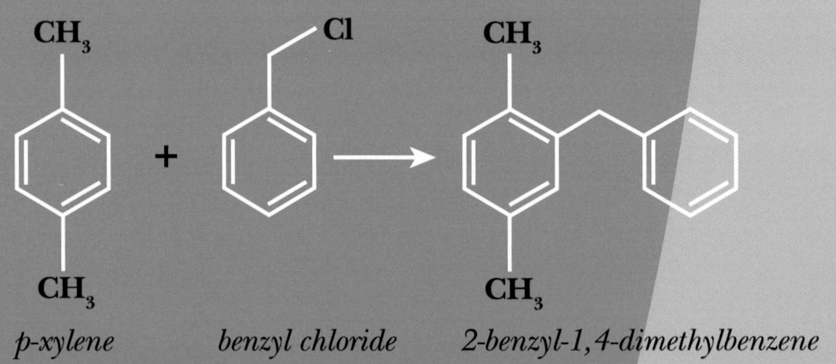

Figure 2: The Friedel-Crafts synthesis of 2-benzyl-1,4-dimethylbenzene

The yield is determined by GC analysis using an external standard. The homogeneous reaction proves to be unselective for the desired product due to trans-alkylation and multi-alkylation reactions. The system also involves an aqueous work-up resulting in the production of a large volume of acidic aluminium containing waste. The heterogeneous catalyst is very selective for the desired product and gives high yields. The work-up involves simple filtration to remove the catalyst, which can be reused.

Learning and discussion points arising from experiment

This practical experiment makes a very obvious case for the use of a heterogeneous Lewis acid catalyst as a substitute for a homogeneous Lewis acid catalyst. In summary the heterogeneous catalyst is superior to the homogeneous catalyst because it:

- Is selective for the desired product
- Gives a superior yield
- Avoids an aqueous quench step as it is readily removed by filtration
- Has a lower E-factor and arguably a lower %AE

In general the student feedback from this experiment has been very positive. Specific learning and discussion points addressed in the experiment include:

- The problems caused by the aqueous work-up and the lack of selectivity in the homogeneous reaction
- The concept of catalyst poisoning for the heterogeneous catalyst – in order to reactivate the catalyst it needs to be heated
- The application of gas chromatography in both quantitative and qualitative analysis

Description of the experiment

The experiment has worked successfully with two groups of students (40 and 22 second year undergraduates respectively over two years). The main problems encountered were:

- The amount of time taken to analyse the yields by GC (12 minutes per sample).
- The fact that the heterogeneous catalyst generally did not work on re-use until the reaction mixture was heated to 100°C.
- The $HCl_{(g)}$ evolved in the reactions attacked the tubing connecting the reaction to the $NaOH_{(aq)}$ scrubber, causing leaks. Acid proof tubing is therefore recommended.

In this practical the students synthesise 2-benzyl-1,4-dimethylbenzene through the Freidel-Crafts catalysed alkylation of *p*-xylene by benzyl chloride in a standard reflux apparatus. No solvent is required in the reaction as *p*-xylene is used in excess. The practical should be carried out in a fume cupboard. Two catalysts have been tested in the reaction, aluminium chloride ($AlCl_3$) and zinc chloride supported on K100 silica. It is anticipated that other Lewis acid catalysts could be used in this reaction. $HCl_{(g)}$ is produced during this experiment and it is recommended that off-gases be passed through a Dreschel bottle containing $NaOH_{(aq)}$. During the experiment the students add benzyl chloride dropwise to a stirred solution of *p*-xylene and catalyst over a period of approximately one hour. The reaction proceeds smoothly if the addition rate is slow and the students can follow the approximate rate based on the frequency of $HCl_{(g)}$ bubbles passing through the $NaOH_{(aq)}$ solution. No external heating is required.

For the homogeneously catalysed reaction an aqueous work-up is required in order to isolate the product and remove the catalyst. For the heterogeneous catalyst, product isolation is facilitated by simple filtration. The recovered catalyst can be reused with fresh reactants; however, sometimes it is necessary to activate the catalyst by heating the reaction to 80°C. Quantitative analysis can be achieved by gas chromatography as described in the Student Guide. The full practical, where homogeneous and heterogeneous Lewis acid catalysts are compared, can be carried out in two five hour laboratory sessions.

There are several possible extensions to this experiment:

Catalyst synthesis and characterisation. In a longer project the students could chose to synthesise the catalyst themselves. This opens up the possibility of catalyst optimisation with respect to loading. It also offers the opportunity for the students to examine the effect of different support materials (e.g. alumina, clay, etc). The students could characterise the catalysts using techniques available in the teaching laboratory.

Synthesis of alternative Lewis acid catalysts. The students could synthesise different supported Lewis acid catalysts, using $AlCl_3$, BF_3, etc. This could form the basis of a mini-project in which the students could investigate the effect of support on the reactivity and selectivity of the Lewis acid.

Different reactions. The students could examine the use of the catalyst in different reaction systems including other alkylations, acylations and Friedel-Crafts catalysed systems. No other reaction systems were tried at York. One particular idea would be to have a multi-step target synthesis of which the alkylation step forms one part. Therefore, a complete synthesis using both traditional and "green" methods could be compared.

3 Clean and efficient synthesis of 4-aminobenzoic acid from 4-nitrotoluene

In this practical the students synthesise 4-aminobenzoic (PABA) acid from 4-nitrotoluene. The practical is aimed at second or third year undergraduate students. The practical involves a two-stage synthesis (Figure 3). In the first step the methyl group is selectively oxidised to a benzoic acid group in a phase transfer catalysis (PTC) reaction. In the second step the nitro group is selectively reduced under mild conditions to the corresponding amine function.

Practical aims

- To synthesise 4-aminobenzoic acid from 4-nitrotoluene
- To carry out selective and mild oxidation of a deactivated methyl benzene to form a benzoic acid in PTC reaction
- To carry out mild and selective reduction on 4-nitrobenzoic acid without reducing the acid group
- To critically assess the pros and cons of each reaction from a Green Chemistry perspective

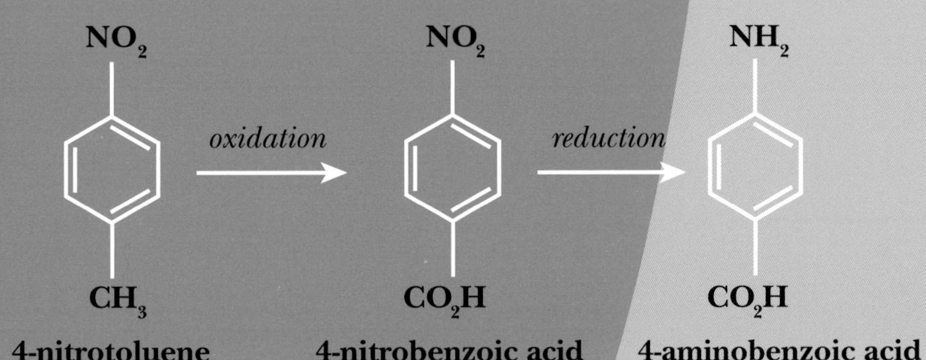

Figure 3: Two stage synthesis of PABA from nitrotoluene

In the second stage the students will have a choice of reaction systems with which to work. All the reactions involve the use of simple, readily available glassware and, in general, mild conditions with respect to temperature and pressure. The experimental yields are good (70 – 90%) and the majority of the chemicals are cheap and readily available.

Learning and discussion points arising from experiment

This experiment offers students hands-on experience with a very efficient and high yielding PTC reaction. It also demonstrates that selective oxidation can be carried out under relatively mild conditions. The PTC oxidation reaction is an excellent mechanism for demonstrating to students the need for careful control (in this case pH) of reaction conditions. In discussion the students are encouraged to explore the relative "greenness" of the PTC reaction in comparison to other oxidation systems as this reaction is far from a clear cut case. Time permitting (a second day in the laboratory will be required) the experiment can be extended to the synthesis of PABA by the selective reduction of the nitro-group. Two methods, adapted from peer reviewed journals, are proposed. This gives the students a choice of methods to use. They should then be encouraged to discuss the relative pros and cons of the chosen system from a green chemistry perspective.

Description of the practical

It should be noted that this practical requires specialised equipment (an organic tolerant pH probe) and an expensive catalyst (ruthenium chloride trihydrate) and therefore may not be appropriate for all teaching laboratories. The reaction should be carried out in a fume cupboard. The apparatus for the oxidation of nitrotoluene consists of a simple two-necked round-bottomed-flask fitted with a dropping funnel and an organic tolerant pH probe. Whilst it is possible to avoid the use of a pH probe by adding NaOH(aq) at a constant rate throughout the experiment, yields were found to be greatly improved if the addition rate was controlled by adjusting the pH. Due to the biphasic nature of the reaction a high stirring rate is recommended to ensure that all components are well contacted. The work-up for the reaction liberates acidic gas and leads to significant foaming and should therefore be carried out slowly. The product is precipitated during the work-up procedure and is readily recovered by filtration. Yields in the oxidation reaction are generally excellent (90% or greater). If time permits, the selective reduction of the nitro-group to the amine can be carried out by one of two suggested methods. Both methods are carried out at room temperature, are relatively straightforward and can be completed comfortably within a five hour laboratory session. No special apparatus is required but the Pd on charcoal catalyst is quite expensive. In both cases the reactions should be carried out in a fume cupboard. The reactions generally proceed with a good yield (>80%).

This experiment has not been tested with undergraduate students to date. It has been tested with a first year PhD student and worked successfully. The key issues associated with the experiment are:

- The cost of the ruthenium and the palladium catalysts.
- The requirement for an organic tolerant pH probe in the PTC reaction.
- The acidic and basic waste generated by the experiment.

For the reduction of 4-nitrobenzoic acid to 4-aminobenzoic acid, the students can either:

a) All follow one specific reduction method; or

b) Be broken into two groups and each group follow a different reduction method

At the end of the practical the students could compare the results of the reduction methods and make some qualitative/quantitative assessment about the relative "Greenness" of each method.

In addition to synthesising the product and calculating yields, there are several other analytical methods that the students could use including:

- Gas chromatography analysis
- Infrared analysis
- Mass spectroscopy
- Nuclear magnetic resonance

They could also be asked to:

a) Draw a mass balanced process flow sheet describing the different processes carried out.

b) Comment on the mass balance of the system and indicate areas of key environmental concern (from both a material and energy perspective).

c) Suggest alternative routes to achieve the synthesis.

d) Undertake a literature search followed by experimental trialling of the synthetic routes (longer term project).

The practical could be extended to a mini-project based on the UV-radiation absorbing properties of PABA. The students could test their PABA product by UV/Vis spectrophotometry. Further analogues of PABA could be synthesised and the effect on the UV-absorbing properties could be assessed.

4

Extraction and conversion of limonene to terephthalic acid via *p*-cymene

This practical is aimed at undergraduate students in their third or fourth year and Masters level students. Different methods of extracting limonene from citrus fruit peel are examined, including steam distillation, solvent extraction and cold pressing. The extracted limonene is then converted to *p*-cymene through one of three methods, and the conversion is followed by gas chromatography.

Practical aims

- To demonstrate a viable alternative route to terephthalic acid from limonene
- To extract and quantify useful chemicals from renewable raw materials
- To convert limonene to *p*-cymene by several methods
- To oxidise *p*-cymene to terephthalic acid using a stoichiometric oxidant
- To critically assess the relative "greenness" of the methods employed in the extraction and synthetic steps

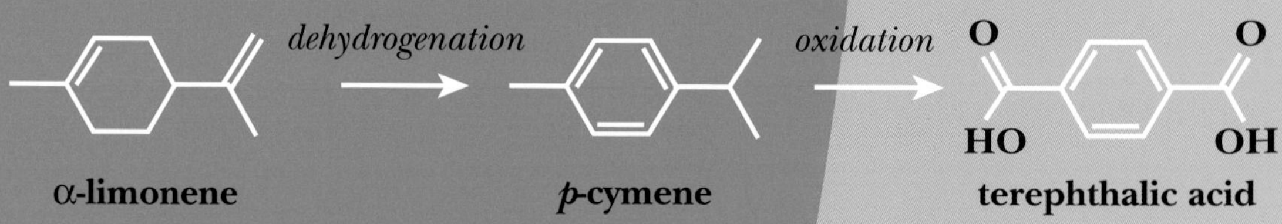

Figure 4: The conversion of limonene to terephthalic acid via *p*-cymene

If time permits, a final oxidation reaction can be carried out to convert *p*-cymene to terephthalic acid. In general, the apparatus required is simple and readily available, however, the experimental set-up for the gas phase conversion of limonene to *p*-cymene requires a small furnace in order to reach a temperature of 300°C. The yields obtained depend upon the experimental methodology followed.

Learning and discussion points arising from experiment

This practical offers the students the opportunity to progress from the primary extraction of a renewable feedstock (limonene) to the production of terephthalic acid, a commodity chemical, in three distinct steps. The practical also offers the students a choice of synthetic method to convert limonene to *p*-cymene. This presents an opportunity for a comparison of the relative "greenness" of the different methods if the cohort of students is of a sufficient size. The students will have the potential to engage with several techniques during this practical, including:

- Different methods of extracting compounds from renewable feedstocks
- Heterogeneous catalysis
- Homogeneous catalysis
- A gas phase continuous flow reactor
- An oxidation reaction

Description of the practical

In general the apparatus used in the practical is simple and should be available in most teaching laboratories, with the possible exception of the steam distillation apparatus and the small furnace required for the gas phase dehydrogenation of limonene. However, neither of these methods is vital for the completion of the practical. The dehydrogenation and oxidation reactions should be carried out in a fume cupboard as should the solvent extraction. In the first part of the practical the students extract limonene from grated citrus fruit peel by either cold pressing, solvent extraction (hexane) or by steam distillation. The limonene content is determined by gas chromatography. Limonene is converted to *p*-cymene in the second step by one of several methods. *N*-lithioethylenediamine is an effective method and generally results in a good yield (>70%). The method involves the use of lithium metal, a nitrogen atmosphere, elevated temperatures and the evolution of hydrogen gas and should definitely be carried out in a fume cupboard. Product purification requires a work-up procedure that produces alkaline aqueous waste. Both the gas phase and liquid phase catalytic dehydrogenation reactions require the use of palladium supported on carbon, which is an expensive catalyst. The liquid phase reaction catalysed by palladium on charcoal requires a nitrogen atmosphere and elevated temperatures. The gas phase dehydrogenation reaction requires a specialist reactor set up and a small furnace capable of heating the reactor to 300°C. In this reaction the limonene is heated to 165°C and passed over a bed of palladium on charcoal (at 300°C) using nitrogen as the carrier gas. The product is collected in a cold trap. In the final step of the practical, *p*-cymene is oxidised to terephthalic acid using stoichiometric quantities of a chromate oxidising agent in the presence of concentrated sulphuric acid. The reaction rate is controlled by the slow addition of sulphuric acid. The reaction system generates large quantities of chromium contaminated acidic aqueous waste. The yield from the reaction is reasonable (>50%).

The experiment has worked successfully with two groups of Masters level students (groups of 15 and 11 students respectively). There are a few points to note about the experiments:

- The liquid phase dehydrogenation (section 4.7) can be quite vigorous and it is recommended that once the reaction has commenced the students leave it to run in a fume cupboard with the sash drawn fully down.
- The gas phase dehydrogenation reaction (section 4.8) takes quite a long time to set up and a long time to reach the desired reaction temperature. The yield in this experiment is also quite low (<20%). It is does however give students an excellent opportunity to experience a gas phase catalytic reaction.

There are several possible extensions to this experiment:

Other extraction methodologies could be incorporated into the experiment, in particular:

- The students could try several different solvents for the extraction of limonene in order to determine the most selective solvent.
- The students could apply cutting edge technology by extracting the limonene using the supercritical CO_2 extraction method developed by Hutchison *et al.*(4)

The method of producing terephthalic acid (TA) is far from green as detailed in this practical. The students could be assigned a project to look into the clean oxidation of *p*-cymene to TA. One potential route is the method described by Sasson *et al*,(5) as detailed in practical 3.

The students could extend the practical by polymerising TA into polyethylene terephthalate (PET).

Summary

The practicals presented here and in the Student Guide are good, but far from perfect, examples of green chemistry. The fact that the practicals are not perfect provides the opportunity to open up valuable student discussion relating to the relative measurement of "greenness" and to experimental improvements. The experiments offer the students much needed experience in several different green chemistry techniques including heterogeneous catalysis, phase transfer catalysis, the use of renewable raw materials, continuous flow reactors and solventless reactions. Students should be further encouraged to critically examine their experiments and calculate a series of green chemistry metrics. By doing this, students become aware of the wider implications associated with chemical reactions and also develop the tools required to compare different synthetic methodologies for a common target. In general the student feedback regarding these practical experiments has been good and many constructive points have been raised.

References:

[1] Constable, D.J.C.; Curzons, A.D., and Cunningham, V.L., Metrics to 'green' chemistry - which are the best? *Green Chemistry.* **2002**, 4, 521.

[2] Price, P.M.; Clark, J.H., and Macquarrie, D.J., Modified silicas for clean technology. *Journal of the Chemical Society-Dalton Transactions.* **2000**, 101.

[3] Clark, J.H.; Cullen, S.R.; Barlow, S.J., and Bastock, T.W., Environmentally Friendly Chemistry Using Supported Reagent Catalysts - Structure-Property Relationships for Clayzic. *Journal of the Chemical Society-Perkin Transactions 2.* **1994**, 1117

[4] L. C. McKenzie, J. E. Thompson, R. Sullivan and J. E. Hutchison, Green chemical processing in the teaching laboratory: a convenient liquid CO2 extraction of natural products. *Green Chem.,* **6**(**8**), (2004), 355-358

[5] Y. Sasson, G. D. Zappi and R. Neumann, Liquid-phase oxidation of deactivated methylbenzenes by aqueous sodium hypochlorite catalysed by ruthenium salts under phase-transfer catalytic conditions. *J. Org. Chem.,* **51**, (1986), 2880-2883